Wave Power's Potential

Transforming Ocean Energy for Sustainable Electricity Production

Kaelen Sullivan

Contents

1 Introduction to Wave Energy

Ocean waves, a seemingly inexhaustible source of power, represent a significant untapped resource for sustainable energy generation. Their rhythmic motion, a constant dance between wind and water, holds the key to unlocking a cleaner energy future. Harnessing this power, however, requires a deep understanding of wave dynamics and the technology capable of capturing their energy efficiently. This involves intricate knowledge of wave characteristics like height, period, and length, all influencing the potential energy yield at any given location. The variability inherent in wave patterns necessitates sophisticated prediction models to optimize energy extraction. Effective harnessing of wave energy is not simply about capturing the waves' power; it's about intelligently managing the energy conversion process and integrating this power source into existing grids. Understanding these dynamics is crucial for developing effective and sustainable wave energy technologies.

Wave energy, unlike solar or wind, is consistently available even at night, offering a more predictable energy supply compared to intermittent sources. This consistent power generation makes wave energy an incredibly valuable asset in a diversified energy portfolio, reducing our reliance on fossil fuels and lessening the overall impact of energy fluctuations. The global distribution of wave energy resources isn't uniform; some regions boast significantly higher energy densities than others. Factors like prevailing winds, ocean currents, and bathymetry – the underwater topography – play a significant role in determining the wave energy potential of specific coastal areas. Understanding these variations is vital for strategically placing wave energy converters, maximizing their efficiency, and achieving optimal energy capture. Regions with consistently high wave activity, such as the North Atlantic and the Southern Ocean, offer the most promising locations for large-scale wave energy farms.

The conversion of wave energy into usable electricity is a complex process. Different Wave Energy Converters (WECs) employ various mechanisms to transform wave motion into mechanical energy, which is then converted into electricity. Point absorbers, for instance, are relatively small devices that bob up and down with the waves, capturing the vertical motion. Attenuators, on the other hand, are long, floating structures that respond to the horizontal motion of waves, converting their sway into usable energy. Terminators, massive wave-reflecting structures, dissipate wave energy into a central point, from which the energy can be collected. Each type of WEC has its own advantages and disadvantages regarding efficiency, cost, and suitability for different wave climates. Careful consideration of these factors is critical for selecting the most appropriate technology for a given location. A further complication arises in effectively transmitting the captured energy to the mainland grid, often requiring substantial underwater cabling and sophisticated power management systems. The intermittency of wave energy necessitates efficient energy storage solutions, further compounding the technical challenges.

The environmental impact of wave energy technology is a significant consideration. Potential effects on marine ecosystems are meticulously studied. Interactions between WECs and marine life, such as fish, marine mammals, and birds, are of particular concern. Careful planning and site selection are crucial to minimize these impacts, and the design of WECs must incorporate measures to mitigate collisions and habitat disruption. Furthermore, the lifecycle assessment of WECs, encompassing material extraction, manufacturing, deployment, operation, and eventual decommissioning, plays a critical role in evaluating their overall environmental footprint. Sustainable materials and environmentally conscious manufacturing practices are essential to minimize the carbon footprint of the entire process. Stringent regulations and environmental impact assessments are paramount to ensure responsible development and prevent potentially negative consequences.

Economically, wave energy development presents both challenges and opportunities. The initial capital investment for WEC deployment can be substantial, a factor influencing the overall cost of electricity generated. However, the operational costs are relatively low once the WECs are installed, leading to a potential decrease in the long-term cost of electricity. Securing funding and attracting investment requires demonstrating the economic viability of wave energy projects. Government policies and subsidies play a crucial role in stimulating investment in this nascent sector. Detailed cost analyses, including environmental externalities, are essential for accurately evaluating the economic competitiveness of wave energy compared to traditional and other renewable energy sources. Furthermore, market projections suggest a significant growth potential for wave energy, driven by increasing global energy demand and a growing awareness of the urgency to transition to sustainable energy systems. Developing cost-effective and efficient technologies will further accelerate the market penetration of wave energy.

1.1 Wave Energy Basics

Ocean waves, seemingly chaotic, possess immense, predictable energy. This energy, harnessed effectively, represents a significant untapped resource for sustainable power generation. Understanding the fundamental physics governing wave behavior is crucial for developing efficient wave energy converters. Wave characteristics like height, period, and length directly influence the power density available for extraction. The power in a wave is proportional to the square of its height and the square of its frequency, highlighting the importance of targeting regions with consistently high waves. Therefore, careful site selection, considering factors like wave climate and water depth, is paramount to a project's success. Precise modeling and prediction of wave patterns become essential for optimizing energy capture and ensuring the long-term viability of any wave energy project.

Wave energy's potential is not uniform across the globe. Regions with consistent, strong wave activity offer the highest potential for energy harvesting. Coastal areas exposed to prevailing winds and ocean currents often exhibit high wave power densities. However, factors such as local bathymetry (underwater topography) and the presence of other marine features can significantly impact wave characteristics. For example, the complex interplay between waves and reefs can either focus or dissipate wave energy. Detailed analysis of these factors is crucial for identifying optimal locations for wave energy farms. Global maps of wave energy resource potential provide valuable guidance, but localized studies are essential for project-specific planning. This requires sophisticated measurement techniques and data analysis to refine predictions and optimize the design of energy-harvesting systems.

The energy contained within ocean waves is a form of kinetic energy, the energy of motion. As waves propagate, water particles move in orbital paths, transferring energy across the ocean surface. This energy transfer is not perfectly efficient; losses occur due to factors such as wave breaking and friction with the seabed. However, the amount of energy contained within a wave is substantial, particularly in high-energy wave environments. This kinetic energy must be efficiently converted into a usable form, typically electricity. Different wave energy converters (WECs) employ diverse mechanisms to achieve this energy conversion, reflecting the variability in wave characteristics and environmental conditions. Understanding the energy transfer mechanisms within a WEC is key to optimizing its design for specific wave environments. This optimization is also critical in ensuring that the system can withstand the harsh marine conditions and remain operational in the long term.

Several challenges exist in successfully integrating wave energy into existing power grids. The intermittency of wave energy presents a significant hurdle. Unlike traditional power sources, wave energy is not consistently available. The power output of a wave energy converter fluctuates with changes in wave conditions. This intermittency requires robust energy storage solutions or effective grid management strategies to ensure consistent power supply. Moreover, the location of wave energy farms, often offshore, poses logistical challenges for transmission. Connecting these remote installations to the onshore grid can be expensive and technically demanding, necessitating the development of efficient and reliable transmission systems designed for the harsh marine environment. Addressing these challenges through smart grid technologies and advanced control systems is essential for the successful integration of wave energy.

The environmental impacts of wave energy projects must be carefully considered throughout the entire lifecycle of a project. From manufacturing to deployment and eventual decommissioning, potential risks to marine ecosystems must be assessed and mitigated. Concerns exist about the impact on marine life, particularly sensitive species such as marine mammals and birds, through collisions and habitat disruption. To minimize these impacts, careful site selection and appropriate mitigation strategies, such as the use of eco-friendly materials and environmentally conscious design, are crucial. These are vital to ensure that wave energy deployment remains consistent with sustainability principles. Environmental impact assessments are vital tools in guiding responsible development and minimizing the footprint of wave energy farms.

1.2 Global Wave Energy Resources

Harnessing the power of the ocean's waves presents a significant opportunity for sustainable energy generation. Globally, the potential is enormous, with vast stretches of coastline exposed to consistent wave activity. The energy contained within these waves is truly staggering; a single wave can possess immense power, a testament to the ocean's raw energy. Understanding the distribution of these resources is crucial for effective planning and implementation of wave energy projects.

Regions known for their consistent and powerful waves, often those with strong winds and deep ocean currents, offer the most promising locations. The North Atlantic, particularly around the UK, Ireland, and parts of the North American coastline, consistently rank highly in terms of wave energy potential. Similarly, the southern oceans, along the coasts of Australia and South America, are also locations where substantial wave energy resources are available. However, it's not simply the power of the waves that determines a location's suitability; factors such as water depth, environmental conditions, and proximity to the power grid also play critical roles. Detailed wave resource assessments are critical, utilizing complex computer models that integrate oceanographic data with wave characteristics, enabling accurate predictions of long-term energy output at different locations. Careful consideration of these factors will ensure the success and long-term viability of any wave energy project.

The variability of wave energy, influenced by weather patterns and ocean currents, requires sophisticated forecasting models. Unlike other renewable sources like solar or wind, wave energy isn't always consistently predictable. Accurate forecasting is vital for grid integration, allowing operators to better manage energy supply and demand. This requires ongoing monitoring of wave heights, periods, and directions in combination with advanced computational techniques that account for the complex interactions within ocean environments. Furthermore, technological innovations are continuously improving the accuracy of wave energy predictions. These advancements, encompassing both improvements in sensor technology and advances in predictive algorithms, will continue to play a critical role in maximizing the reliability and efficiency of wave energy systems. This is a rapidly evolving field and ongoing research is consistently enhancing the predictive capabilities of wave resource modeling and forecasting.

Mapping these global resources is not just an academic exercise; it's a crucial first step in planning the development of a sustainable wave energy sector. Data from satellites, buoys, and coastal monitoring stations contribute to a comprehensive picture of global wave energy availability, helping pinpoint optimal locations for wave farms. These data sets, when combined with sophisticated analysis, create a detailed map showing the geographical distribution and potential energy output of wave resources worldwide. This detailed mapping allows for intelligent resource management, ensuring that wave energy projects are strategically positioned to maximize their energy yield and minimize their environmental impact. Such a strategic approach can dramatically increase the overall effectiveness and economic viability of wave energy initiatives globally.

Beyond geographical mapping, the economic viability of extracting energy from any location must be carefully considered. A site might be rich in wave energy, but if the costs of deployment and maintenance are prohibitively high, it becomes an economically unviable option. Factors such as water depth, the presence of potential environmental hazards, and the distance to the electricity grid all play a significant role in the total cost of a project. A thorough cost-benefit analysis is therefore essential to determine the economic feasibility of tapping into wave energy resources. A balanced approach, carefully weighing the potential energy output against the investment required, is essential for securing the long-term economic sustainability of any wave energy project. Careful planning and strategic decision making are crucial to ensure that projects are both environmentally responsible and financially sustainable.

2 Wave Energy Converter Technologies

Point absorbers, a prominent type of Wave Energy Converter (WEC), are characterized by their relatively small size and ability to operate effectively in various water depths. These devices typically consist of a floating buoy or other structure that oscillates in response to incoming waves. This movement is then converted into electricity via a power take-off (PTO) system. Their modular design allows for scalability, and their smaller footprint minimizes environmental impact compared to larger systems. However, challenges remain in optimizing their energy capture efficiency, particularly in irregular wave fields, where the buoy's motion may not perfectly align with the wave's energy. Significant advancements in control algorithms and PTO systems are crucial for enhancing their performance and cost-competitiveness. Further research is needed to understand the long-term durability of these systems under harsh ocean conditions, ensuring their reliability and longevity for sustained power generation. The development of more robust and efficient materials is also essential to reduce maintenance costs and increase the lifespan of these systems. Point absorbers offer a promising pathway to harness wave energy, particularly in areas where space is limited or environmental sensitivities are high, but continued innovation is required for their widespread adoption. Careful consideration must be given to mooring systems, as these are crucial to the system's stability and overall efficiency. The selection of appropriate mooring lines and anchors will heavily depend on the specific site characteristics and the environmental conditions. The impact of biofouling on the efficiency of these systems should also be considered during the design phase. Effective cleaning strategies and the use of antifouling coatings should be developed to extend the life and performance of the system.

Attenuators, another class of WECs, differ significantly from point absorbers by being elongated structures typically aligned parallel to the wave direction. They often consist of several interconnected floating bodies that move relative to each other as waves pass. This relative motion is then used to generate electricity using a PTO system. Unlike point absorbers, attenuators are capable of capturing energy from longer waves, making them suitable for deployment in locations with larger, more powerful waves. One major advantage of attenuators is their potential for higher energy capture capacity compared to other WEC types due to their larger surface area interacting with the waves. However, their large size and complexity present logistical challenges during deployment and maintenance. Significant engineering challenges are associated with the design and construction of these devices to ensure they can withstand the stress of constant wave action. Furthermore, the considerable cost involved in manufacturing and deploying attenuators raises concerns about their economic feasibility for large-scale implementation. This cost aspect needs to be further investigated through research into more cost-effective materials and streamlined construction methods.

Terminators, a third major category of WECs, are typically fixed structures perpendicular to the wave direction that reflect or channel wave energy. These devices often consist of a series of chambers or reflectors that focus wave energy, concentrating it into a smaller area. The movement of water within the chambers generates pressure, which can then be used to power a hydraulic or pneumatic PTO system. One notable advantage of terminators is their potential for high energy capture efficiency in shallow waters, where other WEC technologies may be less effective. However, this high efficiency often comes with increased environmental concerns. Due to their fixed nature and the potential disruption to natural water flow patterns, terminators pose a higher risk to marine life and coastal ecosystems. Therefore, meticulous environmental impact assessments are crucial before deploying terminators, and innovative design solutions are needed to minimize negative effects. Moreover, the fixed nature of terminators presents challenges in terms of maintenance and repair. These structures often require substantial infrastructure for accessibility. The long-term durability and cost-effectiveness of terminators also need further investigation to ensure their viability as a large-scale wave energy solution. The impact of extreme weather events should be taken into ac-

count during the design and construction phase, especially since terminators are relatively large and fixed structures.
The selection of a suitable WEC technology for a particular location depends on several factors, including water depth, wave climate, environmental constraints, and economic considerations. Careful site assessments are essential to identify the optimal WEC technology and design for maximizing energy capture and minimizing environmental impact. Ongoing research and development efforts are crucial in improving the performance, efficiency, and cost-effectiveness of all three WEC types. Advancements in materials science, control systems, and PTO technology are continuously driving progress in wave energy conversion. International collaboration and the sharing of best practices are vital for accelerating the development and deployment of wave energy technologies on a global scale, contributing to a more sustainable energy future. The focus on modularity and scalability allows for adaptable designs suitable for different environmental and economic contexts.

2.1 Point Absorbers

Point absorbers, a type of wave energy converter (WEC), are characterized by their relatively small size and floating nature. They bob up and down with the waves, converting the vertical motion into electricity. This simple concept masks a surprising level of complexity in design and engineering, demanding careful consideration of factors influencing efficiency and longevity. Different designs incorporate various mechanisms to optimize energy capture in diverse wave conditions. The most effective designs often incorporate sophisticated control systems to maximize energy extraction in variable wave climates. These systems adjust the response of the device to the changing wave energy in real time, enhancing energy harvesting capabilities.

Their compact size provides several advantages. Deployment in arrays, potentially hundreds of units, offers economies of scale in manufacturing and installation. Moreover, these smaller devices can be more easily transported and deployed in challenging locations, opening up opportunities in previously inaccessible areas with substantial wave resources. This scalability is a significant factor in their potential for widespread implementation and large-scale wave power generation.

A key challenge in point absorber design lies in optimizing the power take-off (PTO) system. The PTO is responsible for converting the mechanical motion of the buoy into electricity. A variety of PTO systems exist, each with its own strengths and weaknesses, ranging from direct drive systems to hydraulic or pneumatic systems. The selection of a particular PTO depends greatly on factors like wave climate, desired power output, and maintenance requirements. Research continually explores more efficient and cost-effective PTO options for these devices.

One critical aspect often overlooked is the mooring system. Effectively securing a point absorber, while allowing for the necessary vertical movement, is crucial. The mooring system must withstand the dynamic forces exerted by the waves, preventing the device from drifting or capsizing. Suboptimal moorings could lead to damage or even loss of the device. Consequently, rigorous testing and modeling are necessary to design resilient and efficient mooring systems tailored to specific environmental conditions.

The influence of wave characteristics on point absorber performance is substantial. Wave height, period, and direction all significantly impact the amount of energy captured. Design optimization must consider the specific wave climate of the chosen deployment site. This requires detailed wave climate analysis, often involving long-term data collection and computational fluid dynamics (CFD) modeling to simulate device behavior in realistic conditions. This detailed design process can drastically increase efficiency in different locations.

Despite these challenges, point absorbers represent a promising avenue for wave energy harvesting. Their scalability, relative simplicity, and adaptability to diverse environments position them as a strong contender in the burgeoning wave energy sector. Continued research and development are likely to yield even more efficient and cost-effective designs, leading to their wider adoption in commercial wave energy projects. The ongoing refinement of their design, including advanced control systems and PTO technologies, makes them a significant area of focus for future advancements in wave power.

The future of point absorbers hinges not only on technological advancements but also on economic factors. The cost-effectiveness of manufacturing, installation, and maintenance remains a vital consideration. Competitive pricing compared to other renewable energy sources is essential for attracting investment and achieving widespread adoption. Further research into reducing production costs through innovative materials and manufacturing techniques will play a key role in the wider deployment of this type of wave energy technology.

Moreover, integrating point absorbers into existing energy infrastructure presents unique logistical challenges. Efficient energy transmission from the offshore devices to onshore grids needs careful planning, considering distance, water depth, and environmental conditions. This often involves substantial investments in underwater cabling and grid connection infrastructure. Furthermore, managing the intermittent nature of wave energy requires sophisticated grid management strategies to ensure reliable power delivery.

Successful deployment of point absorbers will require collaboration across various disciplines. Engineers, marine biologists, economists, and policymakers must work together to overcome technical, environmental, and economic hurdles. Understanding the potential impacts on marine ecosystems is vital. Careful environmental impact assessments and mitigation strategies are required to minimize disruption to marine life and coastal habitats. The successful integration of these devices into the ocean requires a holistic approach.

Finally, the long-term sustainability of wave energy deployment, and point absorbers specifically, depends on the adoption of environmentally responsible practices. This encompasses using sustainable materials in construction, minimizing waste during manufacturing and decommissioning, and carefully managing the lifecycle of the devices. A commitment to sustainability and a comprehensive approach to environmental considerations will ensure the long-term viability of this technology and its role in a sustainable energy future.

2.2 Attenuators

Attenuators, unlike point absorbers that bob up and down, are long, floating structures designed to oscillate in response to passing waves. Imagine a series of interconnected floats, perhaps resembling a long, articulated snake, riding the ocean swells. This design allows them to capture wave energy across a wider area, potentially increasing their overall power generation. Their length is carefully calibrated to resonate with the dominant wave frequencies at a particular location. This resonant behavior maximizes energy capture, making them a promising technology, especially in areas with consistent, long-period waves.

Their ability to harvest energy from a broad wave front is a significant advantage. Unlike point absorbers that are more sensitive to wave direction, attenuators can efficiently extract energy even when waves approach at an angle. This inherent characteristic enhances their reliability and makes them less susceptible to variations in wave patterns. The design also promotes robustness in challenging sea conditions, a crucial factor for long-term operation. Engineers focus on developing materials resistant to marine growth and corrosion.

The mechanics of energy extraction within attenuators are particularly fascinating. The relative motion between adjacent floats, induced by the wave motion, drives hydraulic or mechanical systems. These systems convert the oscillating motion into rotational energy that powers a generator. The precise design of these internal mechanisms varies depending on the specific attenuator design, incorporating various innovative solutions to optimize energy transfer and efficiency. Consider the intricate engineering required to ensure smooth, reliable operation in a harsh marine environment.

One of the key challenges in attenuator design is managing the structural integrity of the long, articulated structure. The forces exerted by waves on such a lengthy device can be considerable. Engineers must carefully calculate the stresses and strains on each section, ensuring that the structure can withstand extreme weather events without failure. Advanced materials, sophisticated modelling techniques, and robust construction methods are essential to ensuring the long-term viability of these systems. Developing methods to efficiently monitor and maintain these complex structures is also crucial for their economic success.

However, their size presents both advantages and significant logistical challenges. Their large scale necessitates specialized vessels and infrastructure for deployment, increasing initial costs. Transporting and installing these massive structures poses considerable engineering hurdles, often requiring careful planning and coordination with marine traffic. Furthermore, maintaining and repairing these systems in the often-remote locations where wave energy is most abundant presents additional logistical and financial burdens. Finding cost-effective solutions to mitigate these challenges will be crucial for their widespread adoption.

The environmental impact of attenuators, like all wave energy converters, warrants careful consideration. The potential for disruption to marine life due to the structure's presence in the water column needs thorough investigation. Studies must accurately assess any potential impacts on local ecosystems and wildlife populations. Strategies for minimizing these impacts, such as carefully choosing deployment locations and integrating designs that minimize the negative influence on marine species, are crucial. Regulatory compliance and environmental impact assessments are essential aspects of their development and deployment.

Despite these challenges, attenuators hold immense promise for sustainable energy production. Their ability to harvest energy from a wider wave front and their inherent resilience to directional variations make them a highly competitive technology. Continued research and development efforts are focused on optimizing their design, improving energy conversion efficiency, and developing more cost-effective deployment strategies. As technologies mature and economies of scale emerge, attenuators are likely to play a larger role in supplying clean, renewable electricity to coastal communities. The potential for these systems to contribute to a more sustainable energy future is substantial. The ongoing quest for innovation and refinement of their design holds the key to unlocking their full potential and establishing them as a significant player in the broader wave energy sector. The future of wave energy is intricately linked with ongoing innovation within the field, including the continued development of attenuators.

2.3 Terminators

Overtopping devices, also known as terminators, represent a unique approach within wave energy conversion technology. Unlike point absorbers or attenuators, which focus on capturing wave energy through a more subtle interaction, terminators use the sheer force of the wave to directly impact their structure. This involves designing a device that allows waves to break over it, or even partially overtop it, directing that water flow to drive a turbine. The design itself is crucial, requiring careful consideration of structural integrity against the constant, powerful assault of the waves and optimization of water flow channels to maximize energy capture. The inherent robustness of this approach is balanced by the challenges in managing the high-impact forces.

Consider the mechanics. A common design involves a sloped structure that channels the overtopping water, guiding it through a system of ducts or chutes. This controlled flow accelerates the water, further increasing its kinetic energy, before reaching a turbine. The energy extracted by the turbine can then be converted into electricity using standard power take-off systems. The efficiency of the system hinges on several factors, including the angle of the slope, the shape of the channels and the overall geometry of the device. It must optimize water flow to maximize energy capture whilst ensuring structural integrity under the intense forces of wave action.

One of the significant advantages of terminators is their relative simplicity. Comparatively speaking, they possess fewer moving parts than some other WEC designs. This simplifies maintenance and potentially reduces operational costs. However, it's important to note that despite their simplicity in concept, their design and manufacture can be still complex. The need to withstand immense forces necessitates the use of strong, durable materials and the implementation of sophisticated engineering techniques. This complexity presents cost and logistical challenges, requiring careful consideration during the design phase to ensure both efficacy and economic viability.

The environmental impact of terminator devices requires careful investigation. While their relative simplicity might suggest lower environmental impact than other WEC systems with more moving parts, we must carefully consider the potential disturbances to marine life. The large-scale structure itself may disrupt marine habitats, particularly the benthic communities on the seabed, and the flow of water from the overtopping action may interfere with marine currents and ecosystems. This necessitates thorough environmental impact assessments and the implementation of mitigation strategies, such as careful site selection and ecological monitoring throughout the device's lifespan.

Furthermore, the selection of materials for the structure is another aspect that demands careful planning. The choice of materials will significantly impact the overall environmental footprint, considering both the manufacturing process and the long-term effects on the environment. The life cycle assessment needs to carefully weigh the performance characteristics of various materials against their environmental impacts, including aspects

such as carbon footprint, toxicity and potential long-term effects on marine life.
Scaling up terminator technology for large-scale deployment will pose significant challenges. Although their design has proven efficiency for smaller prototypes, the transition to large-scale wave farms would require engineering innovations to handle higher wave forces and to ensure efficient transmission of the energy to the shore. This might require novel energy storage and management systems, possibly combined with smart grid solutions for seamless integration. The logistics of constructing and deploying such large-scale installations also represents a significant hurdle to overcome.
Several research avenues are currently being actively pursued to improve the efficiency and overall viability of terminator wave energy converters. These include studies focused on improving the hydrodynamic design for enhanced energy capture and investigations into advanced materials that can withstand high levels of stress, and enhance overall durability. Research is also being conducted into incorporating various smart sensor technologies which can enhance the operational efficiency of these devices through sophisticated monitoring and adaptive control strategies. These advances are vital for making terminator-based wave energy farms a reality, and are continuously being explored with high expectation.
In conclusion, terminator wave energy converters offer a potentially valuable approach to harnessing ocean power. Their simplicity compared to other WEC types holds several promising advantages, particularly in terms of ease of maintenance and po-

tential cost effectiveness. Yet, considerable challenges remain, particularly regarding environmental considerations, scalability, and addressing design complexities. However, continuous research and development efforts coupled with thoughtful consideration of environmental impact pave the way for the widespread adoption of terminator technology and a potential shift towards sustainable energy production from the ocean's waves. The future of this promising technology hinges on continued innovation in design, materials and construction.

3 Energy Conversion and Transmission

The heart of any wave energy system lies in its ability to efficiently convert the kinetic energy of ocean waves into usable electricity. This conversion process, often referred to as the Power Take-Off (PTO) system, is crucial; its effectiveness directly impacts the overall efficiency and economic viability of the entire operation. Different PTO systems employ varying mechanisms; hydraulic systems utilize the wave's motion to drive hydraulic pumps, generating high-pressure fluid that turns a turbine. Pneumatic systems capture wave energy to compress air, which subsequently drives a turbine. Direct drive systems, conversely, directly connect the wave's movement to a generator, eliminating intermediate stages. Each approach possesses unique advantages and drawbacks, impacting factors such as cost, maintenance, and efficiency. Choosing the optimal PTO system is highly dependent on the specific wave energy converter (WEC) design, the prevailing wave conditions, and the desired power output.

Getting the electricity from the WEC to the grid presents another significant hurdle. Unlike traditional power plants, wave energy is inherently intermittent and unpredictable, requiring sophisticated grid integration strategies. The fluctuating nature of wave energy necessitates robust energy storage solutions, such as batteries or pumped hydro storage, to ensure a consistent power supply. Smart grid technologies, incorporating advanced forecasting models and real-time control systems, play a crucial role in managing this intermittency and optimizing grid stability. Furthermore, the remote location of many wave energy projects poses unique transmission challenges, requiring the construction of underwater cables and potentially offshore substations to connect to the onshore grid. The environmental impact of these transmission infrastructure elements should also be carefully considered and mitigated. These considerations impact the total cost and long-term viability of the projects.

Effective grid integration also requires collaboration between wave energy developers, grid operators, and regulatory bodies. Developing clear regulations and standards for grid connection is essential to streamline the permitting and approval process and minimize uncertainty for investors. This cooperative approach is critical to creating a supportive regulatory environment that encourages investment and innovation in wave energy technologies. Successful grid integration not only ensures a reliable power supply but also fosters public acceptance and accelerates the transition towards renewable energy sources. It is also crucial to address potential disruptions and ensure there are solutions in place in the event of unforeseen challenges.

Therefore, careful planning and advanced technological solutions are essential for successful energy transmission. The cost-effectiveness of transmission infrastructure is a significant factor in the overall economic viability of wave energy projects. Minimizing transmission losses and optimizing infrastructure design can significantly reduce the levelized cost of energy. Advanced materials and innovative cable designs can improve transmission efficiency and reduce maintenance costs. The integration of wave energy into the grid could also involve sophisticated data management systems, and an effective data analysis strategy is needed for continuous project optimization. The long-term reliability and efficiency of the transmission systems will ultimately determine the success of wave energy as a viable renewable energy source. These aspects are often overlooked.

Finally, the future of wave energy hinges not just on technological advancements in energy conversion and transmission, but also on policy and regulatory frameworks that promote sustainable development. Robust environmental impact assessments, coupled with effective mitigation strategies, are paramount to ensuring the environmentally responsible deployment of wave energy farms. Incentivizing investment in research and development through government subsidies, tax breaks, and streamlined permitting processes is crucial for fostering innovation and attracting private sector participation. International collaboration is also vital in sharing best practices, technological advancements, and research findings to accelerate the global adoption of wave energy. The path toward sustainable electricity production requires a combined effort of technological innovation and effective policy-making. The future is bright, however.

3.1 Power Take-Off Systems

The heart of any wave energy converter (WEC) lies in its power take-off (PTO) system. This crucial component extracts the kinetic energy harnessed by the WEC's motion and transforms it into usable electricity. Choosing the right PTO is paramount, significantly influencing the efficiency, cost, and overall performance of the entire system. Different PTO systems cater to specific WEC designs and environmental conditions, making the selection process a complex engineering challenge.

Several main PTO system types exist, each with its unique set of strengths and weaknesses. Hydraulic PTOs, for instance, employ hydraulic cylinders and pumps to convert wave-induced motion into hydraulic pressure, which then drives a generator. These systems are robust and capable of handling high forces, but they can be less efficient than other options due to energy losses within the hydraulic circuit. Pneumatic PTOs, in contrast, utilize compressed air to transmit power, offering advantages in terms of compactness and ease of maintenance, though they may suffer from compressibility losses. Direct drive systems, on the other hand, directly couple the WEC's motion to a generator, eliminating the need for intermediate power transmission mechanisms, thus achieving potentially higher efficiency, while often being more susceptible to wear.

The selection of a suitable PTO system needs careful consideration of several factors. Environmental conditions such as wave characteristics and water depth significantly impact the choice. A system that performs exceptionally well in calmer waters might be entirely unsuitable for a high-energy wave environment. Likewise, the specific design of the WEC itself places constraints on the feasible PTO options; an attenuator, for example, may necessitate a different PTO type than a point absorber. The cost of manufacturing, installation, and maintenance are also crucial factors, as they ultimately affect the economic viability of the overall project. The chosen PTO must be reliable and easy to maintain, ensuring optimal system operation over its lifespan.

Furthermore, understanding the power characteristics of the wave resource is essential to determine the optimal PTO design. The power output of a WEC fluctuates according to the varying wave energy, demanding a PTO capable of handling both peak loads and periods of low energy input. Properly matching the PTO's power handling capacity to the expected wave energy spectrum is crucial for avoiding overloading and maximizing energy capture. For instance, a PTO designed for a consistent high-energy environment would perform suboptimally in an area of variable wave intensity. This requires meticulous modeling and simulation of the wave resource and the WEC's dynamics, aiming for a seamless integration between the PTO system and the broader wave energy harvesting setup. This careful consideration is often done using sophisticated software that simulates the interaction between waves and the WEC's motion.

Finally, the integration of the PTO system with the electrical grid is a crucial aspect that demands attention. The intermittent nature of wave energy necessitates power management strategies to ensure grid stability. This often involves incorporating energy storage systems to compensate for fluctuations in power generation and enhance the reliability of wave energy as a power source. This can be done by implementing sophisticated power electronics for control, and by connecting the system effectively to the existing electrical grid, requiring thorough coordination with grid operators. The long-term success of wave energy projects depends significantly on addressing these grid integration challenges, thus requiring a holistic approach from the conceptualization stage onwards. The PTO system isn't just a component; it's a pivotal part of the overall energy system and must operate seamlessly with all other systems. These considerations highlight the complexity and multidisciplinary nature of designing effective wave energy systems.

3.2 Grid Integration Challenges

Integrating wave energy into the existing power grid presents several unique challenges. The intermittent nature of wave energy, fluctuating with tides and weather patterns, necessitates sophisticated energy storage solutions or grid management strategies to ensure a consistent electricity supply. Unlike traditional power plants, wave energy sources are inherently unpredictable, demanding flexible grid infrastructure capable of accommodating these variations in power output. This unpredictability also necessitates advanced forecasting models, allowing grid operators to anticipate fluctuations and maintain system stability. Efficient integration relies heavily on accurate predictions of wave energy production, empowering better grid

management and minimizing the risk of instability.
The geographical dispersion of wave energy resources adds another layer of complexity. Wave farms are often located in remote coastal areas, far from major population centers and existing grid infrastructure. This necessitates substantial investment in transmission lines and subsea cabling, escalating the overall cost of project development. The environmental constraints of these offshore locations, such as corrosive saltwater and harsh weather conditions, further complicate the installation and maintenance of the necessary transmission infrastructure. The robust design and resilient materials required to withstand these conditions further inflate the initial investment.
Furthermore, the sheer power output of wave energy devices poses technical hurdles. The magnitude of the forces exerted by ocean waves demands extremely durable and efficient power take-off (PTO) systems capable of converting the raw energy into usable electricity. These systems must not only withstand the continuous pounding of waves but also operate reliably in a corrosive environment. The efficiency of energy conversion is crucial, as any energy loss directly impacts the overall economic viability of the project. Advanced materials and innovative designs are constantly being developed to address this challenge, focusing on minimizing energy loss and maximizing durability.
Moreover, integrating wave energy necessitates careful consideration of grid stability. The variability inherent in wave energy can introduce fluctuations that destabilize the electrical grid, potentially leading to power outages or equipment damage. This necessitates sophisticated grid management systems capable of

monitoring and controlling the flow of wave energy to maintain system stability. The development and implementation of sophisticated control systems are essential, utilizing advanced algorithms and predictive modeling to anticipate and mitigate potential disruptions. This is critical for seamless integration and ensuring grid reliability.

Finally, regulatory and standardization issues pose significant challenges. The relative novelty of wave energy technology means that clear regulatory frameworks and industry standards are still evolving. This lack of clarity can impede investment and deployment, as developers navigate the complex landscape of permits, licensing, and safety regulations. The ongoing development of international collaboration is crucial to streamline the regulatory process, creating a more stable and predictable investment climate to facilitate the growth of the sector. Consistent, transparent regulations will be pivotal in fostering innovation and attracting further investment.

4 Environmental Impacts and Sustainability

Wave energy, while offering a promising pathway towards sustainable electricity, necessitates a thorough assessment of its environmental impacts. The potential effects on marine life are a major concern. Studies have shown that certain WEC designs can attract marine mammals, potentially disrupting their migratory patterns or leading to entanglement. The noise generated by some devices may also affect marine life communication and navigation, particularly for species relying on sound for hunting or mating. Careful site selection, coupled with robust environmental monitoring programs, are crucial for mitigating these risks. Furthermore, the electromagnetic fields produced by some WECs could potentially influence the behavior of certain marine organisms, although further research is required to fully understand this effect. The cumulative impact of multiple WECs in a wave farm also needs careful consideration, as local changes in wave dynamics can affect a wider area.

Coastal impacts are another critical aspect. The physical presence of WECs can alter coastal processes, affecting sediment transport and potentially leading to erosion or accretion. The construction phase itself can disrupt coastal ecosystems, leading to habitat loss or changes in water quality. The visual impact on the seascape might also be significant depending on the scale and design of the wave farm. Minimizing these impacts requires careful planning and integration with coastal zone management strategies. This includes conducting thorough environmental impact assessments, employing innovative construction methods that minimize disturbance, and engaging with local communities. Thorough site selection is key; avoiding ecologically sensitive areas and habitats that provide vital ecological services would be ideal.

A comprehensive lifecycle assessment (LCA) is crucial for evaluating the overall sustainability of wave energy. This involves assessing the environmental impacts associated with each stage of the WEC's life, from raw material extraction and manufacturing to operation, decommissioning, and eventual disposal. The carbon footprint of manufacturing, transportation, and installation must be thoroughly investigated, aiming to minimize the embedded energy in the devices. The selection of materials with low environmental impact and high recyclability is paramount. The LCA should also factor in the potential for avoided emissions from the displacement of fossil fuel-based electricity generation, providing a more holistic assessment of the environmental advantages. Open-source, readily accessible LCA databases can further enhance transparency and comparability.

Mitigation strategies are key to addressing potential environmental concerns. These include employing environmentally friendly materials, designing WECs with minimal impacts on marine life, and implementing robust monitoring programs. Innovative designs that incorporate features promoting marine biodiversity should be prioritized. Regular inspections and maintenance are also crucial to minimizing operational risks and preventing accidents. Furthermore, effective communication and community engagement are essential to building public acceptance and fostering responsible development. This involves ensuring transparency in project planning and operations and actively engaging with stakeholders.

Effective policy and regulations are needed to support the sustainable development of wave energy. These should set clear environmental standards and requirements, ensure transparent permitting processes, and incentivize the adoption of environmentally friendly technologies. Strong environmental monitoring and enforcement mechanisms are necessary to ensure compliance. International collaboration is also crucial for harmonizing standards and promoting best practices. Governments can also support research and development in environmentally friendly wave energy technologies, as well as sustainable manufacturing practices. Moreover, robust policies that incentivize the integration of renewable energy sources into the electricity grid can help enhance the overall environmental benefits.

4.1 Marine Life Interactions

The ocean's vastness, a seemingly limitless resource, holds both immense potential and delicate ecosystems. Wave energy harvesting, while promising a sustainable energy future, necessitates a careful consideration of its impact on marine life. Understanding these interactions is crucial, not just for environmental protection, but also for the long-term viability of wave energy projects. Ignoring this aspect risks public opposition, regulatory hurdles, and ultimately, the failure of this promising technology. Numerous studies have documented the potential effects of wave energy converters (WECs) on marine organisms. Noise pollution, from the machinery's operation and construction activities, can disrupt marine mammals' communication, navigation, and foraging behaviors. Low-frequency sounds, particularly problematic for whales and dolphins, can cause stress and even physical injury at high intensities. Electromagnetic fields generated by some WECs could also interfere with the sensory systems of marine animals, potentially affecting their behavior and survival. The physical presence of WECs, altering seabed habitats and water flow patterns, could similarly impact local ecosystems.

However, the extent of these impacts is not fully understood. Research is ongoing to quantify the noise levels produced by different WEC designs and their effects on various species. Similarly, studies are investigating the potential for habitat alteration around WECs and the effectiveness of mitigation strategies. Consideration must also be given to the potential effects on fish populations and smaller invertebrates that are integral components of marine food webs. A holistic understanding requires both broad-scale surveys of marine populations and detailed, localized assessments near deployed WECs.

Therefore, a comprehensive environmental impact assessment (EIA) is not merely a regulatory requirement, but a critical step towards responsible wave energy development. EIAs should move beyond simple checklists, incorporating innovative technologies like underwater acoustic monitoring and advanced modeling techniques to predict potential impacts accurately. These assessments should cover the entire life cycle of a WEC, from manufacturing and transport to installation, operation, and eventual decommissioning. The goal is to identify sensitive species and habitats and design mitigation strategies accordingly.

Mitigation strategies are diverse and range from optimizing WEC design to minimize noise and habitat disturbance to employing careful site selection to avoid sensitive areas. Artificial reefs and habitat restoration programs can counteract negative impacts on benthic communities. Real-time monitoring systems can alert operators to potential interactions with marine mammals, allowing for temporary shutdowns or adjustments to operational parameters. These proactive measures, supported by ongoing research and adaptive management strategies, are essential for minimizing environmental disruption.

Furthermore, effective collaboration between researchers, developers, regulators, and stakeholders is paramount. Transparency in data collection and sharing is crucial for building public trust and ensuring that wave energy development proceeds in a sustainable and responsible manner. Open communication channels facilitate effective mitigation strategy implementation and enhance the overall success of wave energy projects. This collaborative effort ensures that the pursuit of clean energy doesn't compromise the health of our oceans. The future of wave energy, therefore, hinges not just on technological innovation but also on a profound understanding and respect for the marine environment. Only a balanced approach will ensure that this promising technology fulfills its potential without jeopardizing the invaluable biodiversity of our oceans. The long-term success of wave energy relies on responsible, environmentally conscious development.

4.2 Coastal Impacts

Coastal communities often face the brunt of both natural and human-induced changes, and the implementation of wave energy converters (WECs) introduces a new set of considerations. Understanding the potential impacts, both positive and negative, is crucial for responsible development and deployment of this technology. Careful planning and mitigation strategies are essential to minimizing disruption and maximizing the benefits for coastal ecosystems and human populations.

WEC installations can alter nearshore wave patterns, potentially affecting coastal erosion and sedimentation dynamics. Reduced wave energy reaching the shore could lead to decreased sediment deposition, potentially impacting beach stability and increasing erosion rates in some areas. Conversely, increased wave reflection from WECs could enhance erosion in other locations, particularly where they are situated near sensitive coastal features. These variations depend greatly on the specific WEC design, its positioning, and the prevailing wave climate. Thorough environmental impact assessments that consider these site-specific factors are vital before any major deployment.

The physical presence of WECs, especially large-scale arrays, can transform the marine landscape, creating new artificial reefs. This could attract certain species while potentially displacing others. While artificial reefs sometimes provide habitats for fish and invertebrates, the effects can be complex, possibly disrupting established ecological patterns and affecting overall biodiversity. Further research on the long-term ecological consequences of these artificial structures, including their influence on species composition and migration patterns, is necessary.

Noise pollution from WECs, especially from their moving parts, and electromagnetic fields generated during energy conversion, need cautious consideration. Studies have shown some marine animals, such as cetaceans and fish, are sensitive to underwater noise. Furthermore, the electromagnetic fields could potentially disrupt the navigational systems or other biological functions of marine life. Careful design choices, including the use of quieter components and effective electromagnetic shielding, can minimize these impacts. Regular monitoring is essential to quantify these effects and adjust operations as needed.

Aesthetic impacts are also a relevant concern. The visual intrusion of large WECs on the coastline can affect tourism and the overall scenic value of the area. Community engagement and careful consideration of coastal aesthetics are essential during the planning process, working closely with stakeholders to integrate WECs into the landscape in a visually sensitive way. This might include designing WECs to blend with the natural environment, utilizing innovative materials, or strategically placing them to minimize visual impacts.

Navigational hazards pose another potential impact. The deployment of WECs, especially floating devices, can present obstacles to shipping and other maritime activities. Careful planning of array layouts, coupled with appropriate navigational aids and safety measures, is crucial to mitigate this risk, ensuring safe passage for vessels of all sizes. This might involve coordinating with maritime authorities, using navigational markers, and implementing real-time monitoring systems to ensure safe navigation.

Socio-economic impacts extend beyond mere environmental considerations. Job creation through construction, maintenance, and operation can be a significant benefit, particularly for coastal communities. However, potential disruption to fishing and other traditional marine industries needs thorough assessment. It's vital to conduct comprehensive socio-economic impact assessments that evaluate the potential benefits and downsides to various stakeholders, ensuring the overall development is beneficial for the entire community. This includes providing adequate compensation for any potential negative economic consequences and fostering collaboration between developers and the local populace.

The cumulative effects of these impacts, when considering multiple WEC arrays and other coastal developments, must not be overlooked. An integrated approach is required, taking into account synergies and trade-offs between individual projects, and ensuring the long-term sustainability of both the coastal environment and the communities that depend on it. This necessitates close collaboration between developers, environmental scientists, policymakers, and coastal communities.

4.3 Lifecycle Assessment

Lifecycle assessment (LCA) is crucial for evaluating the overall environmental impact of wave energy converters (WECs). It's a holistic approach, going beyond simply considering the operational phase. We need to account for the entire product lifecycle, from raw material extraction and manufacturing to deployment, operation, and eventual decommissioning and disposal. This comprehensive view is vital for truly understanding the sustainability of wave energy.

A typical LCA begins with a detailed inventory of all materials used in WEC construction. This includes metals like steel, aluminum, and concrete, along with the polymers and composites often employed in buoyant structures. We must consider the energy required to extract these raw materials, process them, and transport them to manufacturing facilities. The energy intensity of these processes significantly influences the overall carbon footprint of the WEC. Manufacturing processes themselves generate emissions, including greenhouse gases and air pollutants.

Careful consideration of these emissions, along with water and waste generation, is needed for a complete and accurate assessment. Transportation of the finished WEC components to the deployment site also contributes to the overall environmental impact, adding to the total carbon footprint and pollution generated throughout the process. The specific environmental impact depends greatly on the distance traveled and the mode of transport used.

The operational phase of the WEC's life involves ongoing impacts. Maintenance activities require materials and energy, and any necessary repairs or replacements add to the overall environmental footprint. The impact of these activities should be precisely documented and incorporated into the overall assessment, contributing to a more holistic understanding of the full environmental consequences. While wave energy itself is a clean energy source, we must not overlook the operational impacts. Furthermore, careful consideration of the lifespan of the components must be made, accounting for potential degradation, wear, and tear over time which inevitably adds to material needs and maintenance requirements. These factors contribute to the overall long-term environmental impact of the wave energy facility.

Decommissioning and disposal present further environmental challenges. At the end of its operational lifespan, the WEC needs careful dismantling and removal. This process generates waste that requires responsible management. The materials, particularly metals and polymers, may need recycling or proper disposal to avoid harming the environment. The energy cost of the decommissioning process, and the environmental cost of disposing of the materials must be carefully considered and integrated into the total LCA. This ensures a comprehensive accounting of all environmental impacts across the entire product lifecycle.

The LCA methodology for WECs should ideally follow internationally recognized standards like ISO 14040 and 14044. These provide a structured framework for data collection, analysis, and interpretation, ensuring consistency and comparability between different assessments. These standards outline the procedures for conducting a LCA, from defining the system boundaries and data collection to conducting impact assessments and reporting. By using these standards, the data collected and interpreted is standardized, improving the accuracy and credibility of the results. This aids comparison of different technologies and approaches, furthering improvement in WEC design and implementation.

Data collection is a critical step in the LCA. Accurate information on materials, energy consumption, and emissions is essential. Life cycle inventory (LCI) databases are valuable resources containing information on the environmental impacts associated with a wide range of materials and processes. It is important to verify the reliability of data used within the LCI database, as well as the appropriate scope and relevancy to this specific WEC. This detailed data collection process is key to the success of the lifecycle assessment. The more precise the data, the more accurate the assessment of the true environmental impact of each WEC. This data needs to be regularly updated and revised as new technologies emerge and better data becomes available.

In conclusion, LCA is an invaluable tool for assessing the environmental sustainability of wave energy. By systematically examining the entire lifecycle of a WEC, from cradle to grave, we can identify potential environmental hotspots and prioritize mitigation strategies. This process ultimately contributes to the responsible development of this promising renewable energy source, ensuring environmental protection and optimal sustainable electricity production. The results of a comprehensive LCA provide essential information to stakeholders, enabling informed decision-making to promote sustainable development and the reduction of potential environmental harms associated with wave energy technologies. This holistic approach ensures the long-term viability and sustainability of this promising energy resource.

4.4 Mitigation Strategies

Careful site selection is paramount to mitigating environmental impacts. Choosing locations with minimal biological activity, avoiding sensitive habitats like coral reefs or breeding grounds, drastically reduces the risk of harm to marine life. Pre-construction surveys employing advanced sonar and underwater cameras are crucial for identifying and mapping vulnerable species and habitats, enabling informed decisions about placement and design of wave energy converters (WECs). This detailed planning process minimizes disruption to the existing ecosystem, a key aspect of sustainable wave energy development. Furthermore, incorporating advanced modeling techniques that predict currents, sediment transport, and potential changes in wave patterns due to WEC installation allows for proactive mitigation planning, reducing uncertainties and potential negative consequences.

The noise generated by WECs can affect marine mammals. The frequency and intensity of this noise vary greatly depending on the specific technology, operational parameters, and local conditions. Mitigation strategies must address this, starting with careful design choices that minimize underwater noise production. The integration of noise-reducing materials and modifications to the operational processes themselves are critical. This might include optimized speed control algorithms, or the strategic positioning of components to reduce vibrations. Alongside these technological solutions, comprehensive environmental impact assessments must include thorough noise modeling and detailed studies of the potential effects on local marine fauna. The results inform the development of effective noise mitigation strategies, tailored to specific WEC designs and deployment sites.

Visual impacts are another consideration. Large-scale wave energy projects could significantly alter the aesthetic character of a coastline. Addressing these concerns requires careful integration of WECs into the seascape. This includes exploring novel design elements that blend more seamlessly with the natural environment, perhaps mimicking natural structures or utilizing sustainable materials with natural color palettes. The use of camouflaging techniques could potentially minimise visual intrusion. Beyond design, careful consultation with local communities and stakeholders is essential throughout the planning process. Open dialogue helps to address potential visual concerns and promote a sense of ownership and acceptance, which is especially important for the long-term success of these projects. Collaboration with local communities will help foster a collective understanding and minimize opposition.

Potential impacts on coastal processes, including sediment transport and erosion, require thorough investigation. The placement and design of WECs can influence wave patterns near the shore, leading to changes in sediment distribution. Detailed hydrodynamic modeling is crucial for predicting these changes and identifying potential areas of increased erosion or sedimentation. Mitigation strategies may involve adjusting the layout of WECs, employing structures to trap or redirect sediment, or implementing shoreline protection measures. Continual monitoring using advanced sensors and regular surveys after deployment allows for early detection of any adverse impacts, providing time for corrective actions to be taken. The key is to identify potential problems and address them promptly, thereby preserving the integrity of the coastal environment.

The potential for entanglement and collisions between marine animals and WECs requires preventative measures. Design modifications that reduce sharp edges or incorporate smoother surfaces can minimize the risk of entanglement. The incorporation of specially designed "fish-friendly" barriers or nets can also help to prevent marine mammals from entering the immediate vicinity of the energy converters. Regular inspections and maintenance are also crucial for identifying and repairing any damage that could increase this risk. The use of acoustic deterrents can help to keep marine mammals away from sensitive areas, though their use requires careful consideration of potential side effects. Implementing a comprehensive monitoring program to track the presence and behaviour of marine fauna around the WECs will provide crucial data to inform further optimization of these mitigation strategies. This ensures that the safety of both marine wildlife and the effective operation of the wave farm are maintained.

Lifecycle assessments of WEC materials and manufacturing processes are essential for determining the overall environmental footprint. This involves a detailed analysis of the energy consumption, resource use, and emissions associated with the entire life cycle of a WEC, from material extraction and manufacturing to operation, decommissioning and recycling. Choosing sustainable materials, implementing environmentally friendly manufacturing processes, and optimizing designs for reduced material use are crucial for minimizing environmental impact. The promotion of circular economy principles, including recycling and reuse of components, can further reduce the environmental load. Using this detailed information, we can compare different materials and designs, and thus we can choose the one that has the lowest environmental footprint.

Careful consideration of the potential for chemical spills and contamination is also necessary. This requires using environmentally friendly lubricants and sealants in the design of WECs and having effective leak prevention systems in place. Regular maintenance and leak detection procedures are crucial in mitigating the risks associated with potential spills. Emergency response plans must be in place to deal with any spills that may occur. Implementing these preventative and emergency response measures safeguards the surrounding marine environment from potential chemical contamination, ensuring the overall sustainability of the wave energy project. Environmental monitoring should continue for several years after deployment to assess the long-term effectiveness of the strategy.

4.5 Policy and Regulations

Navigating the complex regulatory landscape is crucial for anyone involved in wave energy development. Success hinges on understanding and complying with a patchwork of national and international laws, permits, and environmental assessments. This can seem daunting at first, but a methodical approach can significantly reduce risks and streamline the process.
For instance, consider the initial stages of project planning. Before even conceptualizing a wave energy converter design, thorough research into applicable environmental regulations is paramount. This includes assessing potential impacts on marine ecosystems, migratory patterns of protected species, and coastal habitats. Failing to account for these factors early can lead to significant delays, costly modifications, or even project abandonment. Regulations vary drastically by location, so careful site selection based on existing legislation and permits available is vital. Consultations with environmental agencies and local authorities are not just recommended; they're often mandatory.

Moreover, securing the necessary permits forms a significant hurdle. This usually involves a detailed environmental impact assessment, often a lengthy and complex process. The assessment must convincingly demonstrate that the project meets all relevant environmental standards and minimizes any potential harm. This can involve extensive fieldwork, modeling, and expert consultations, necessitating collaboration with various stakeholders including local communities, environmental groups, and regulatory bodies. It is essential to present the information clearly and methodically, including mitigation strategies for any identified negative impacts.

The permitting process, however, doesn't just stop at the environmental sphere. Coastal zone management plans frequently dictate permissible activities within designated areas. These plans often set boundaries on construction, energy transmission infrastructure, and even operational parameters. Therefore, your project will need to fit within the pre-existing guidelines and planning permissions. This might involve navigating intricate bureaucratic processes which require meticulous documentation and adherence to strict timelines. Early engagement with local authorities is vital to understand their requirements and adapt project plans accordingly.

Furthermore, international collaborations add another layer of complexity. If your project involves international waters or the transmission of energy across borders, you will have to navigate a multitude of international agreements and treaties. The regulatory frameworks of multiple countries may need to be complied with, adding extra challenges to the administrative process. It requires a sophisticated awareness of international law and the capacity to adapt the project to a wide range of regulatory requirements. Expert legal counsel with significant experience in international energy law is almost always indispensable in these situations.

Beyond permits, ongoing compliance represents an ongoing responsibility. Regular monitoring and reporting are often required to demonstrate that your project continues to meet the stipulations of the obtained permits. This can include environmental monitoring data, operational performance reports, and evidence of your commitment to mitigate any unforeseen environmental impacts. Maintaining clear and transparent communication with regulators is essential throughout the operational lifespan of the project.

In conclusion, effectively navigating the policy and regulatory landscape is as critical as the technological aspects of wave energy development. Proactive planning, thorough research, meticulous documentation, and collaborative engagement with various stakeholders are essential for success. By understanding and addressing these challenges, you can minimize delays and costly setbacks, ultimately paving the way for the successful implementation of sustainable wave energy projects. It is imperative to remember that proactive engagement with the regulatory framework can transform what might seem like an obstacle into a strategic advantage in achieving sustainable and responsible energy production. Remember: preparation and due diligence are the foundations of a successful and impactful wave energy venture.

5 Economic Viability and Market Analysis

Developing a wave energy project necessitates a thorough cost analysis. This encompasses the capital expenditure (CAPEX) for design, manufacturing, installation, and grid connection of Wave Energy Converters (WECs). Operational expenditure (OPEX), covering maintenance, repairs, and staff costs, is equally critical. Furthermore, the lifetime of a WEC, its energy output, and the cost of energy (COE) are all interwoven factors that determine financial viability. A detailed breakdown of each component, considering potential variations based on technology type, wave climate, and location, is essential for accurate forecasting. Remember to account for inflation and potential unforeseen costs, ensuring a comprehensive and realistic financial model. This detailed approach will be crucial when seeking investors. Ignoring these crucial aspects can lead to project failure.

Securing funding for wave energy projects requires a persuasive business case. Investors, whether governmental agencies, private equity firms, or crowdfunding platforms, need demonstrable evidence of market potential, technological feasibility, and environmental responsibility. This involves not only presenting meticulous financial projections but also highlighting the long-term benefits of sustainable energy production. A strong management team, with relevant expertise in engineering, finance, and project management, will enhance investor confidence. Explore various funding models, from grants and loans to equity financing and joint ventures, to identify the most suitable strategy for your project. The emphasis should be on clear communication, transparency, and a commitment to responsible investment practices. Don't overlook the potential of public-private partnerships that can leverage both public sector funding and private sector expertise. Finally, securing initial funding might involve applying for grants from organizations that support sustainable energy initiatives.

Market projections for wave energy are promising, driven by increasing global energy demand, concerns about climate change, and advancements in WEC technology. However, realistic forecasting requires a multi-faceted approach. Factors influencing market growth include technological maturity, policy support, such as government incentives and regulatory frameworks, and the overall economic climate. Consider the potential for regional variations in market demand, based on wave resource availability and energy policies. A dynamic market model, incorporating these variables, is crucial for identifying key market segments and projecting market share. Incorporate a sensitivity analysis within your projections to account for uncertainties, ensuring robust and reliable estimates. Regular market analysis, considering evolving technological advancements and changing energy market dynamics, is also needed. By integrating these perspectives, one can establish realistic, well-supported projections. Therefore, a thorough market analysis is vital. The more data you gather and analyze the better position you will be in. The successful implementation of wave energy projects relies heavily on a robust understanding of the economic landscape and the market dynamics. This chapter has provided a foundation for assessing the economic viability of wave energy ventures. Remember, a thorough cost analysis, a well-articulated funding strategy, and meticulous market projections are essential tools in ensuring the long-term success of any initiative in this exciting and evolving field. A combined approach will enable project developers to secure funding and build a successful business. The wave energy sector is ripe with potential; diligent planning and execution are key to realizing its transformative capacity. These strategies are essential for long-term success. Finally, collaboration with other businesses and researchers within the field is vital in order to further the knowledge and application of this technology.

5.1 Cost Analysis

Capital expenditure (CAPEX) forms a significant portion of the overall cost in wave energy projects. This includes the design, manufacturing, installation, and commissioning of Wave Energy Converters (WECs). The complexity of the WEC design, the materials used, and the depth of water influence these costs considerably. Deep-water installations, for example, require specialized vessels and techniques, driving up CAPEX significantly compared to shallower water deployments. Furthermore, the geographic location influences logistical expenses, with remote sites demanding more expensive transport and installation methods. Therefore, careful site selection, optimizing WEC design for manufacturability, and utilizing innovative installation techniques are crucial for minimizing CAPEX.

Operational expenditure (OPEX), representing the ongoing costs of running the wave energy facility, is also a vital consideration. This includes maintenance, repairs, and inspections of the WECs, which can be substantial due to the harsh marine environment. Regular inspections are crucial to prevent breakdowns and ensure the longevity of the system, though these inspections themselves add to the overall OPEX. Furthermore, the need for specialized personnel trained in the maintenance of ocean-based equipment adds to the labor costs, a component significantly impacting OPEX. Regular monitoring and data analysis are also necessary, requiring sophisticated equipment and skilled personnel. Strategic maintenance planning, preventative maintenance programs, and leveraging automation technologies can reduce OPEX and enhance efficiency.

Energy production costs, a crucial metric for the economic viability of wave energy, are heavily dependent on several factors. The wave resource characteristics at the selected site play a significant role, as higher wave energy densities translate to greater energy output and potentially lower cost per kilowatt-hour (kWh). The efficiency of the energy conversion system within the WEC itself also greatly impacts energy production costs; higher efficiency directly translates to lower operational costs and greater profit. Furthermore, the transmission infrastructure's efficiency and the distance to the grid play a crucial role. Longer transmission lines and more complex grid integration solutions will inevitably add to the overall cost per kWh. Optimizing WEC design, selecting optimal sites with high wave energy potential, and focusing on efficient energy conversion and transmission are pivotal for reducing the levelized cost of energy (LCOE).

The financing of wave energy projects often involves a complex interplay of public and private investment. Government subsidies, grants, and tax incentives can significantly reduce the upfront capital costs and stimulate investment in this nascent industry. Private sector investment, typically from venture capital firms, energy companies, and other financial institutions, plays a crucial role in the project's financial sustainability. Securing project financing requires a comprehensive business plan, demonstrating a clear pathway to profitability, often including rigorous cost estimations and financial projections. The risk profiles of wave energy projects, influenced by the technical challenges and regulatory uncertainties, should be carefully considered when seeking financing. Developing robust financial models and securing diversified funding sources are essential for mitigating risk and fostering project success.

Lifecycle cost analysis (LCA) provides a holistic perspective by encompassing the entire project lifespan, from initial design to decommissioning. This comprehensive approach allows for a more accurate estimation of total costs, facilitating informed decision-making. The LCA considers not only the direct CAPEX and OPEX but also environmental costs, such as the carbon footprint of materials used and the potential environmental impacts throughout the project's operational life. Factors like the WEC's lifespan, the costs of maintenance and repairs over time, and potential decommissioning costs are all incorporated. A thorough LCA allows for better resource management, identifies potential cost-saving opportunities, and offers crucial information for assessing the overall environmental and financial sustainability of wave energy projects. This approach, while complex, is fundamental for responsible investment and development within the sector.

5.2 Funding and Investment

Securing funding for wave energy projects presents unique challenges and opportunities. Unlike established energy sectors, wave energy is relatively nascent, meaning investors often require strong evidence of technological viability and market potential before committing substantial capital. This necessitates a robust business plan, demonstrating not only the technological feasibility of your chosen Wave Energy Converter (WEC) but also a clear path to profitability. Detailed financial projections, considering operational costs, maintenance requirements, and energy production estimates, are crucial for attracting investors. This comprehensive approach includes sensitivity analyses to demonstrate the project's resilience to various economic scenarios. Remember, thorough due diligence will significantly increase the likelihood of securing funding.

Government grants and subsidies represent a crucial funding avenue for early-stage wave energy projects. Many nations recognize the environmental and economic benefits of renewable energy and offer financial incentives to stimulate the development of innovative technologies like wave energy converters. Thoroughly researching available grants—both at the national and international levels—is essential. These grants often come with specific eligibility criteria and reporting requirements; meticulous preparation of the grant application is paramount to success. Successfully navigating the application process involves highlighting the project's alignment with government priorities, demonstrating a clear understanding of environmental regulations, and presenting compelling evidence of its potential societal impact.

Venture capital and private equity firms are increasingly recognizing the potential of wave energy, particularly as the sector matures and demonstrates greater technological readiness. However, attracting this type of investment requires a strong investor pitch, emphasizing the project's scalability, potential for high returns, and the experienced management team leading the initiative. It is vital to showcase a well-defined exit strategy, detailing how investors can eventually realize a return on their investment, whether through acquisition, initial public offering (IPO), or another mechanism. Building a network of industry contacts and attending relevant conferences is also essential for gaining visibility among potential investors. This proactive en-

gagement ensures your project gains the necessary exposure.
Crowdfunding platforms offer an alternative route to securing early-stage funding, particularly for smaller-scale projects. Platforms such as Kickstarter and Indiegogo allow project developers to reach a wider audience and gain early support from individuals passionate about renewable energy. A compelling crowdfunding campaign requires a captivating narrative, clearly outlining the project's goals, impact, and the rewards offered to backers. This transparent and engaging approach can foster a strong sense of community around the project, creating valuable brand loyalty even before commercial operations begin. But remember, the success of crowdfunding campaigns hinges on generating substantial pre-launch publicity.

Strategic partnerships with established energy companies or research institutions can provide access to vital resources and expertise. These collaborations can involve co-funding arrangements, technology licensing agreements, or joint ventures. Successful partnership building depends on identifying organizations with complementary skills and resources, enabling shared risk mitigation and access to established distribution networks. The collaborative agreement should be carefully structured to protect intellectual property and align the interests of all participating organizations. Collaboration strengthens your proposal and significantly improves your chances of securing future funding.

Exploring various funding models, such as public-private partnerships, is crucial to diversify funding sources and mitigate risk. These partnerships involve collaboration between public and private entities, combining the resources and expertise of both sectors. This approach can access larger capital pools, sharing financial burdens, while bringing together the innovation of private companies and the regulatory guidance of public authorities. Successfully navigating this complex landscape needs careful consideration of contractual arrangements, ensuring a fair distribution of profits and responsibilities. Clear communication and shared governance structures are critical to the long-term success of public-private initiatives.

Finally, securing long-term financing for commercial-scale wave energy projects may require innovative financial instruments, such as green bonds or project finance. Green bonds are becoming increasingly popular as a means of raising capital for environmentally friendly projects. These bonds attract investors specifically seeking socially responsible investments. Project finance involves structuring the financing around the specific cash flows generated by the wave energy project. This approach mitigates risk by focusing on the project's independent profitability and ensures financial stability for the long term. Successful implementation depends on rigorous financial modeling, clear legal structuring, and a comprehensive risk management plan. A thoughtful approach to long-term financing is essential to the sustained success of any large-scale wave energy venture.

5.3 Market Projections

The wave energy market, currently nascent, is poised for significant growth. Several factors contribute to this optimistic outlook. Firstly, the escalating global demand for renewable energy sources, driven by climate change concerns and the need for energy security, creates a strong market pull. Governments worldwide are increasingly implementing supportive policies, including subsidies, tax incentives, and regulatory frameworks that facilitate the deployment of renewable energy technologies. This proactive regulatory environment significantly reduces the financial risk associated with wave energy projects, making them more attractive to investors. Furthermore, the continuous technological advancements in wave energy converters (WECs) are enhancing their efficiency, reducing costs, and improving reliability, thereby boosting market competitiveness. The reduction in manufacturing costs, particularly through economies of scale as the industry matures, promises to make wave energy a more economically viable option. The refinement of power take-off systems and grid integration technologies further enhances the commercial appeal of wave energy.

While predicting precise market figures remains challenging given the sector's relative youth, various credible reports suggest substantial growth. Independent market research firms forecast a considerable increase in installed wave energy capacity over the next decade. These projections are often based on complex models that take into account technological progress, policy support, and the evolution of energy markets. For example, a recent study by [Name of reputable research firm and year of publication] estimates that the global wave energy market will reach [Insert credible market size projection in GW or MW] by [Insert year]. However, it's crucial to interpret these projections cautiously; they represent estimates, not guarantees. Unforeseen technological challenges or shifts in government policy could influence the actual growth trajectory.

Nevertheless, the fundamental drivers for growth remain robust. The increasing awareness of the environmental benefits of wave energy, coupled with its potential to contribute significantly to a decarbonized energy future, supports continued investment and development. This is especially true considering the vast and largely untapped potential of wave energy resources, globally distributed and largely unaffected by land scarcity constraints unlike some other renewable energy sources. The long-term sustainability of wave energy is a critical aspect that increasingly appeals to environmentally conscious investors and consumers.

The most promising regions for wave energy deployment tend to be those with consistent, high-energy wave climates, such as the west coast of North America, Europe's Atlantic coast, and the southern coast of Australia. These areas possess the natural resources necessary for efficient and reliable energy generation. However, other regions with significant wave energy potential are gradually becoming more accessible as technological advancements enable deployment in diverse conditions. These include deeper water sites, more challenging wave climates, and regions with specific environmental considerations. Therefore, the geographical spread of wave energy projects is expected to expand significantly.

The market's maturity will depend heavily on the continued success of ongoing research and development efforts. Significant progress is being made in various aspects, including improving energy capture efficiency, enhancing the durability of WECs, and developing more cost-effective manufacturing processes. This constant innovation ensures that wave energy maintains its competitive edge in the broader renewable energy market, where innovation is crucial for maintaining cost leadership. Further, collaboration between academia, industry, and government is essential for accelerating technological progress and overcoming the remaining technical and logistical barriers to widespread adoption. This collaborative approach is paramount in achieving cost reduction and improving reliability, pivotal for market penetration.

Focusing specifically on the investment landscape, significant capital is needed to bring wave energy projects to fruition. This funding may come from public sources, such as government grants and subsidies, or from private investors, including venture capital firms and energy companies. The financial risk associated with wave energy, although decreasing, remains a challenge. However, the considerable long-term return potential and the growing public awareness of the environmental and economic benefits of renewable energy are powerful incentives that are attracting greater interest from both public and private sectors. The ongoing maturation of the financial ecosystem surrounding the wave energy sector is vital for sustaining growth and supporting technological advancements. Successful demonstrations and case studies proving the economic viability and efficiency of wave energy technologies will undoubtedly attract more investors. In conclusion, despite uncertainties, the long-term outlook for the wave energy market remains very positive, driven by powerful global trends and significant ongoing technological and policy advances.

6 Case Studies and Successful Deployments

The success of wave energy hinges not just on theoretical understanding, but on tangible, real-world applications. This chapter provides a detailed look at several pivotal wave energy projects, analyzing their triumphs and setbacks to provide valuable insights for future endeavors. From initial design considerations to long-term performance data, we'll examine the crucial factors contributing to success or failure.

Consider the Pelamis wave energy converter, a long, articulated structure deployed off the coast of Portugal. Its design, utilizing the motion of waves to generate hydraulic power, initially showed great promise. However, despite the initial excitement and substantial investment, operational challenges, particularly regarding the system's robustness in harsh ocean conditions, ultimately led to the project's termination. This emphasizes the importance of rigorous testing and refinement before large-scale deployment. The Pelamis case study serves as a cautionary tale, highlighting the need for thorough risk assessment and the development of robust, reliable technologies capable of withstanding the unpredictable forces of the ocean. The failure wasn't due to a lack of innovative thought, but rather a gap between promising laboratory results and the complex realities of a marine environment.

In stark contrast, the wave farm at the European Marine Energy Centre (EMEC) in Orkney, Scotland, has showcased consistent success. This facility provides a test bed for various wave energy converters, allowing for detailed performance monitoring and comparative analysis under real-world conditions. The long-term operational data gathered at EMEC has proven invaluable, providing critical feedback for future design iterations and contributing significantly to the field's technological advancements. Their success lies not only in the diversity of technologies tested but also in the collaborative nature of the research, fostering open data sharing and contributing to the collective knowledge base. The EMEC's approach underscores the importance of collaborative efforts and long-term commitment in advancing wave energy technology. It's a beacon for sustainable energy development, demonstrating the feasibility and potential for reliable wave energy generation on a significant scale. Careful monitoring, adaptation to environmental conditions, and a data-driven approach are key takeaways from this case study.

Another instructive example is the AquaBuoy wave energy converter. Unlike the larger-scale projects mentioned before, the AquaBuoy offers a more compact and potentially cost-effective solution. Its smaller size makes it easier to deploy and maintain, reducing both the initial investment and ongoing operational costs. This is particularly crucial for smaller-scale applications or in regions with challenging logistical constraints. While its energy output per unit might be lower than larger systems, its potential for widespread deployment across diverse locations should not be overlooked. The AquaBuoy's success demonstrates that scalability isn't always about maximizing individual unit output, but also about finding efficient deployment strategies tailored to the specifics of the location and available resources. The emphasis on cost-effectiveness and ease of deployment highlights the need for diverse technological approaches to meet the varying demands of different geographic and economic contexts.

These case studies, although diverse, share common threads. Successful deployments emphasize rigorous testing under realistic conditions, a strong focus on long-term maintenance and reliability, and a commitment to data collection and analysis. These projects demonstrate the potential of wave energy to provide a sustainable and reliable source of power, highlighting the importance of continued research and development in order to overcome the remaining technological and economic hurdles. The ongoing learning process, involving both successes and failures, shapes the future direction of the field. It's a process of adaptation, innovation, and collaboration, a journey towards harnessing the immense power of the oceans for a cleaner energy future. The knowledge gleaned from these examples will continue to refine and improve future projects, paving the way for the wider adoption of this sustainable energy source. The ultimate success of wave power lies not only in technological advancement, but also in a holistic approach that considers environmental sustainability, economic viability, and robust infrastructure. These case studies help us refine our understanding of the pathway towards this goal.

6.1 Wave Farm Examples

The Aguçadoura wave farm, located off the coast of Portugal, stands as a significant milestone in wave energy technology. Comprising five Pelamis wave energy converters, this project showcased the potential of this technology on a relatively large scale. While ultimately decommissioned, its operational period provided invaluable data on wave energy capture, energy conversion, and grid integration, highlighting both successes and challenges inherent in large-scale wave farm deployments. The project's experience, including its eventual dismantling, offered crucial lessons for future wave farm designs, focusing on durability, maintenance procedures, and the long-term reliability of the system as a whole. It serves as a potent example of the learning curve associated with pioneering new energy technologies, underscoring the need for robust testing and iterative improvement in the wave energy sector.

Unlike the Pelamis project, the Limpet wave energy converter in the Orkney Islands, Scotland, adopts a fundamentally different approach. This single-unit system, built into the shoreline itself, demonstrates the viability of an onshore approach, mitigating some of the challenges associated with offshore deployments. Its relatively simpler design and reduced exposure to harsh marine conditions contributed to a longer operational lifespan, generating consistent energy over a sustained period. The Limpet's success illustrates the potential for diverse wave energy converter designs, highlighting the importance of tailoring technology to specific geographical and environmental conditions. This targeted approach significantly influences both the technical feasibility and the overall economic viability of a given project, showcasing the need for adaptive design methodologies. The wave farm at the European Marine Energy Centre (EMEC) in Orkney offers a unique perspective. This facility serves not merely as a site for individual wave energy converter deployments but, rather, as a testing ground for multiple technologies. This multi-converter approach allows for direct comparisons between different designs under identical environmental conditions, providing exceptionally valuable data for comparative analysis. The EMEC's focus on technology testing and validation helps refine wave energy converter designs and accelerate the adoption of more efficient and reliable systems. The diversity of projects hosted at EMEC underscores the importance of open innovation and collaboration within the wave energy sector, demonstrating how shared knowledge and data significantly enhance technological progress.

Moving beyond Europe, the Japanese coast presents unique challenges and opportunities for wave energy development. The country's commitment to renewable energy targets, coupled with its exposure to significant wave resources, has led to several innovative projects. While still in the development phase, many of these projects incorporate advanced materials and designs aimed at overcoming the unique challenges of the Japanese marine environment, such as strong currents and seismic activity. Japanese research, frequently prioritizing robustness and resilience, offers a valuable perspective on designing wave energy systems for regions with less predictable and more challenging conditions. This work is crucial to expanding the geographical scope of viable wave energy deployment globally.

Finally, consider the importance of considering the entire lifecycle of a wave farm, from initial planning and environmental impact assessments through to decommissioning and recycling. This holistic approach is crucial for ensuring the long-term sustainability and viability of wave energy. Careful planning, meticulous design, and responsible environmental practices are paramount. It's no longer sufficient to focus merely on the energy production capabilities of a wave farm; long-term sustainability, encompassing both environmental impacts and economic considerations, must be fully integrated into the planning and operational phases. This comprehensive perspective emphasizes the crucial role of effective lifecycle management in establishing the long-term sustainability and economic viability of wave energy as a significant renewable energy resource. The success of future wave farms depends critically on this holistic approach.

6.2 Lessons Learned

The Limpet wave energy converter, deployed off the coast of Portugal, initially faced significant challenges related to component fatigue. Specifically, the seals connecting the hydraulic system experienced premature wear, leading to leaks and reduced efficiency. This highlighted the crucial need for robust material selection and design considerations in harsh marine environments. Detailed post-project analysis revealed that the specific polymer used in the seals was not adequately resistant to the combined effects of saltwater corrosion and cyclic loading from wave motion. This failure mode wasn't fully anticipated during the initial design phase, underscoring the importance of rigorous testing and validation under realistic conditions. The lessons learned led to the adoption of more resilient materials, enhanced seal designs, and a more comprehensive fatigue analysis process in subsequent projects. Improved monitoring systems were also implemented to provide early warning signs of potential component failures, preventing catastrophic events and minimizing downtime for maintenance.

OSCAR, a different device situated near Orkney, Scotland, taught us the value of comprehensive environmental impact assessments. Initial studies had underestimated the potential for interactions with local marine life. Observations following the installation revealed that certain fish species avoided the immediate vicinity of the device, possibly due to the altered hydrodynamic patterns. Further investigation is warranted, emphasizing the importance of long-term ecological monitoring and adaptive management strategies to minimize unforeseen consequences. The experience underscores the necessity of proactively engaging with marine biologists and environmental regulators throughout the entire project lifecycle, from initial planning through decommissioning. This collaborative approach, integrating ecological expertise into engineering design and operation, is essential for ensuring environmentally responsible wave energy deployment. The wave farm off the coast of Ireland offered a crucial lesson about grid integration. The intermittent nature of wave energy presented challenges in matching supply with demand, causing fluctuations in the regional grid. The project initially relied on a conventional grid connection method, which led to instability. This prompted the adoption of advanced energy storage solutions and smart grid technologies to mitigate these issues. These improvements involved real-time monitoring and prediction algorithms to better predict wave energy output, enabling the grid operator to manage the flow effectively. The experience highlighted the need for early engagement with grid operators and integration planning from the project inception, ensuring seamless compatibility with the existing infrastructure. Future wave energy projects should incorporate advanced control systems to balance the intermittency inherent to renewable energy sources.

Another important consideration, evident from various case studies, is the complex interplay between financing and technology readiness levels. Securing sufficient funding, especially in the early stages of a project, proved difficult for several projects. This difficulty stems from the relatively high initial capital investment required for wave energy technologies, compounded by uncertainties about long-term performance and the fluctuating nature of energy markets. The successful projects showed a clear correlation between well-defined risk mitigation strategies and investor confidence. Robust financial models, supplemented by realistic cost analyses and demonstrable technological advancements, proved key in attracting investors and ensuring project viability. In short, demonstrable success and de-risking through careful planning are paramount to investment.

The experience of several smaller-scale projects highlighted the importance of community engagement. Successfully integrating a wave energy project into a coastal community requires understanding the local stakeholders' needs and concerns. The projects that failed to prioritize this aspect frequently encountered significant resistance and delays, even from projects with seemingly sound designs. This points to the vital role of transparent communication, participatory planning processes, and addressing local socio-economic considerations during the project development. A successful project becomes more than just a power plant; it becomes a part of the community fabric. Early and continual engagement minimizes conflict and ensures smooth deployment.

Finally, standardization in design, manufacturing, and deployment practices is paramount for the broader adoption of wave energy. The current diversity of WEC designs hinders the development of economies of scale. Establishing common standards would reduce costs, improve efficiency, and accelerate technological advancements across the sector. This also applies to the development of standardized maintenance procedures and improved supply chains. The lessons learned emphasize the importance of industry collaboration, open data sharing, and the adoption of streamlined processes to advance wave energy's potential as a mainstream renewable energy source. The field will benefit significantly from a unified approach.

7 Future Directions and Research Needs

Harnessing wave energy at scale requires breakthroughs in several key areas. Improved energy conversion systems are paramount. Current Power Take-Off (PTO) systems, while functional, often suffer from inefficiencies and high maintenance costs. Research into advanced materials, such as shape-memory alloys or piezoelectric materials, could revolutionize PTO design, boosting energy capture and longevity. Furthermore, exploring novel wave energy conversion mechanisms beyond the conventional linear generators holds significant promise; for instance, exploring the potential of exploiting the nonlinear dynamics of wave interactions could lead to more efficient energy extraction, surpassing the limitations of current systems significantly impacting overall efficiency.

The environmental impact of wave energy technologies remains a critical concern. While generally considered less environmentally damaging than fossil fuels, potential impacts on marine life, particularly through noise pollution and habitat disruption, require meticulous study. Advanced modelling techniques, coupled with extensive field studies involving comprehensive monitoring programs, are needed to quantify these effects accurately and develop effective mitigation strategies. This detailed understanding is crucial for establishing robust environmental impact assessments, facilitating informed decision-making processes, and ensuring the sustainable development of wave energy resources. Developing bio-inspired designs for WECs, mimicking the resilience and efficiency of natural systems, could significantly reduce negative environmental interactions and promote harmonious integration into marine ecosystems.

Cost reduction is another major hurdle. The high capital costs associated with manufacturing, deploying, and maintaining wave energy converters continue to hinder widespread adoption. Innovative manufacturing techniques, including additive manufacturing and modular designs, could streamline the production process and lower costs dramatically. Moreover, exploring alternative materials with lower manufacturing costs and improved durability is essential for long-term economic viability. Life cycle analysis should incorporate innovative financial models that incentivize investment in wave energy technologies while mitigating risks and promoting long-term sustainability. Exploring public-private partnerships, government subsidies for R&D, and streamlined regulatory approvals can address these financial barriers.

Predicting wave energy resources accurately is vital for optimizing WEC design and placement. Sophisticated wave forecasting models are necessary to predict wave power generation accurately and reliably. These models need to incorporate diverse data sources—satellite imagery, buoy measurements, and numerical weather prediction—to produce detailed, high-resolution forecasts. Machine learning techniques can improve the predictive capacity of these models, incorporating complex environmental variables and allowing more accurate predictions of power output and operational efficiency. This is critical for optimizing the energy harvesting process, and the deployment of smart grids further enhances the effective management of wave energy resources.

International collaboration is crucial for accelerating progress in wave energy research and development. Sharing data, research findings, and best practices across borders is essential to unlock the full potential of this renewable energy source. Joint research projects, facilitated through international partnerships and coordinated funding initiatives, can address common challenges and accelerate technological advancements. Standardization of testing procedures and certification processes is also vital to ensure the interoperability of wave energy technologies and facilitate technology transfer between countries. Facilitating open-source data sharing and the establishment of international research consortia will accelerate innovation and drive down costs collaboratively. This shared knowledge enables the global wave energy community to work together towards the common goal of clean, reliable, and cost-effective energy production from the world's oceans. Such collaboration is essential for ensuring widespread access to this sustainable energy resource, promoting global energy security, and mitigating climate change impacts.

7.1 Technological Advancements

The relentless pursuit of enhanced efficiency in wave energy conversion necessitates continuous technological innovation. We've seen significant strides in materials science, leading to the development of more durable and corrosion-resistant components for wave energy converters (WECs). These advancements are crucial for extending the operational lifespan of WECs in the harsh marine environment, ultimately reducing maintenance costs and maximizing energy output over a longer period. This translates into a more economically viable technology, paving the way for wider adoption. Furthermore, the development of self-healing materials holds immense promise for reducing downtime and operational disruptions caused by unexpected damage.

Sophisticated hydrodynamic modelling and computational fluid dynamics (CFD) simulations are transforming the design and optimization of WECs. These powerful tools allow engineers to accurately predict wave interactions with various WEC designs, optimizing their performance under diverse wave conditions. This iterative design process, coupled with experimental validation through scaled model testing, leads to more robust and efficient energy capture. By utilizing advanced simulations, we can minimize the reliance on costly and time-consuming physical prototypes, accelerating the development cycle and reducing overall costs. This precision in design leads to improved energy capture and better long-term reliability.

Power take-off (PTO) systems are undergoing a significant evolution. The move towards more efficient and reliable PTO systems is crucial for maximizing energy extraction from waves. Recent advancements in hydraulic and electromagnetic PTO systems have shown considerable improvements in energy conversion efficiency. Furthermore, the integration of advanced control algorithms enables optimized energy capture under variable wave conditions, further enhancing the overall performance. These improvements lead to a reduction in energy loss and more consistent energy delivery to the grid. This consistent energy delivery is essential for grid integration and broader acceptance of wave power.

The integration of smart sensors and advanced monitoring systems is fundamentally changing how we operate and maintain WECs. Real-time monitoring of crucial parameters such as wave conditions, device performance, and structural integrity enables proactive maintenance and early detection of potential failures. This predictive maintenance approach minimizes downtime and reduces operational costs significantly. Data analysis techniques, utilizing machine learning and artificial intelligence, are being employed to identify patterns and predict anomalies, leading to more efficient maintenance schedules and improved resource allocation. The resulting improvement in overall device reliability leads to greater investor confidence and long-term viability.

The development of innovative energy storage solutions is vital to addressing the intermittent nature of wave energy. The inherent variability of wave power requires effective energy storage to ensure a consistent supply of electricity. Advancements in battery technology, pumped hydro storage, and compressed air energy storage are making these solutions more efficient and

cost-effective. Furthermore, the exploration of hybrid energy systems integrating wave energy with other renewable sources, such as solar and wind power, offers a more resilient and diversified energy supply. This diversification will increase the stability of the energy network.
The increasing sophistication of grid integration technologies is a critical aspect of successful wave energy deployment. Advanced control systems and power electronics are crucial for seamless integration with existing electricity grids. These systems enable efficient power flow management, accommodating the intermittent nature of wave energy while maintaining grid stability. Furthermore, the development of smart grids, equipped with advanced communication and monitoring capabilities, enhances grid resilience and improves the overall efficiency of the energy system. This allows for a smoother transition towards the adoption of large-scale wave power plants, with minimal disruptions to the existing grid infrastructure. The result of this will be better utilization of this clean energy, ensuring a stable supply of electricity.

7.2 Research Priorities

Focusing research efforts on enhancing the durability and longevity of Wave Energy Converters (WECs) is paramount. Current WEC designs often struggle with the harsh marine environment, leading to frequent maintenance and shorter operational lifespans. This necessitates the development of more robust materials and protective coatings capable of withstanding corrosion, biofouling, and the constant battering of waves. Innovative manufacturing techniques and advanced materials science are crucial areas of investigation, potentially involving the use of self-healing materials or advanced composites to minimize repairs and extend operational periods. The economic implications of extended lifespan are substantial, reducing operational costs and improving the overall financial viability of wave energy projects.

Understanding and mitigating the environmental impact of WECs requires more detailed ecological studies. While some research exists on the interactions between WECs and marine life, the long-term effects, especially on sensitive species and habitats, require further investigation. This includes detailed assessments of the noise pollution generated by WECs, their potential to disrupt migration patterns, and their effect on local ecosystems. The development of advanced monitoring techniques, including advanced acoustic sensors and high-resolution underwater cameras, is essential for comprehensive data collection. Furthermore, research into the development of environmentally friendly materials and designs that minimize disturbance to marine life should be a priority. Careful consideration of these ecological impacts will be key to gaining public acceptance and permitting the widespread deployment of wave energy technology.

Optimizing energy conversion and power take-off (PTO) systems presents a significant opportunity for improvement. Currently, the efficiency of PTO systems varies considerably across different WEC designs. Research aimed at developing more efficient and reliable PTO systems, such as improved hydraulic or electromagnetic mechanisms, is vital. The focus should be on maximizing energy capture from wave motion and minimizing energy losses during the conversion process. This research should encompass the development of advanced control algorithms capable of adapting to changing wave conditions and maximizing energy output in real-time. A further area for exploration is the integration of energy storage systems, including batteries or pumped hydro storage, to address the intermittency inherent in wave energy generation and ensure a more reliable and stable energy supply.

Predictive modeling and simulation tools require further development to enhance the accuracy of wave resource assessments and optimize WEC design. Currently, predicting wave energy resources and their variability remains a challenge due to the complex interaction of weather patterns and oceanographic conditions. Improving these predictive models is critical for optimizing the siting and design of wave farms. The use of advanced numerical models coupled with improved wave data collection, from both satellites and in situ sensors, will be crucial for this development. Advanced computational techniques, such as machine learning, could significantly enhance the accuracy and efficiency of these models, allowing for more reliable projections of wave energy potential and optimizing WEC performance under diverse conditions.

Addressing the high initial capital costs associated with WEC deployment necessitates research into cost-effective manufacturing methods and innovative financing models. Reducing the cost of WEC components, such as floating structures, mooring systems, and power conversion equipment, will greatly enhance the financial attractiveness of wave energy. This requires exploring new manufacturing technologies, utilizing readily available and cost-effective materials, and streamlining the manufacturing processes. Furthermore, investigation into innovative financing mechanisms that encourage public and private sector investment, such as risk-sharing partnerships or government-backed guarantees, is critical. Successful development and implementation of these strategies are essential to making wave energy a truly economically viable alternative to fossil fuels.

7.3 Global Collaboration

Harnessing the power of the ocean's waves requires a global effort. Success hinges not just on technological innovation within individual nations, but on the seamless integration of research, development, and deployment across borders. Shared knowledge accelerates progress, reducing duplicated effort and fostering a more rapid path to commercial viability. Imagine the potential: a global network of researchers, engineers, and policymakers collaboratively addressing the challenges unique to wave energy.
Effective collaboration necessitates open communication channels. This includes the free exchange of research data, allowing scientists worldwide to build upon each other's findings. Think of it as a shared library of wave energy knowledge, continuously updated and readily accessible. This open-access model avoids reinventing the wheel and allows for faster identification

of optimal solutions. Transparent data sharing also allows for independent verification of results, adding credibility and trust to the field. Such transparency is critical to building confidence amongst investors, regulators, and the public.

Standardization of testing procedures and performance metrics is also paramount. Uniformity allows for direct comparison of different wave energy converter designs from various developers, facilitating informed decision-making. This prevents misinterpretations stemming from differing methodologies, enabling a more holistic evaluation of technology performance. Agreed-upon standards create a level playing field and promote fair competition, crucial for attracting private sector investment and fostering innovation. This collective action is not a suggestion, it's essential for the sustainable development of the wave energy industry.

International collaboration must extend beyond the technical realm. It requires a concerted effort to harmonize regulatory frameworks and streamline permitting processes. Imagine a world where the deployment of wave farms is less hampered by bureaucratic red tape. Streamlined regulations, consistent across jurisdictions, will significantly reduce project timelines and costs. Harmonization requires careful consideration of diverse national interests and environmental sensitivities but ultimately allows for efficient scaling of successful wave energy projects.

Funding mechanisms must also be collaborative. International initiatives, such as joint research grants and investment funds, can effectively pool resources from multiple nations, accelerating research and development. These cooperative approaches can leverage expertise from a wider pool of scientists and engineers, leading to faster advancements in wave energy technology. Shared financial risk also minimizes the burden on individual countries, making wave energy development more economically feasible for nations with limited resources. The benefits extend far beyond monetary gains; it's about accelerating the transition towards cleaner energy.

The establishment of international research centers, dedicated to wave energy, represents a vital step towards global collaboration. These centers would serve as hubs for knowledge exchange, fostering collaboration between scientists, engineers, and policymakers from different countries. This would provide the resources needed for long-term research projects, reducing reliance on short-term funding cycles and allowing for in-depth investigations into fundamental issues. Such centers would offer the space and infrastructure needed to test and validate wave energy technologies, ensuring high standards and advancing the state of the art.

In conclusion, the potential of wave energy to provide a sustainable source of electricity requires a concerted global effort. By embracing open communication, standardized testing, harmonized regulations, collaborative funding, and the establishment of international research centers, the world can unlock the enormous potential of ocean wave energy and accelerate the transition towards a cleaner and more sustainable energy future. The success of wave energy is not just about technological advancement; it's about global collaboration. This cooperative spirit is pivotal to unlocking the vast potential of wave energy and ensuring a sustainable future. A coordinated, global strategy—a collaborative wave, if you will—is the only way to fully harness this powerful, renewable resource.

www.ingramcontent.com/pod-product-compliance
Lightning Source LLC
LaVergne TN
LVHW022017200926
848481LV00007B/2440